A Prairie Year

Messages to Max

by George Rohde

SPECIAL THANKS TO THE FOLLOWING
FOR THEIR HELP IN MAKING THIS BOOK POSSIBLE:

BARB SOLBERG, ERIK FRITZELL, BRUCE LYTLE AND KEVIN ZAVORAL.

ISBN 10: 1-59152-126-2
ISBN 13: 978-1-59152-126-6

Published by Beverly Rohde

For more information, write Farcountry Press, PO Box 5630, Helena, MT 59602

You may order extra copies of this book by calling Farcountry Press toll free at (800) 821-3874.

Produced by Sweetgrass Books
PO Box 5630
Helena, MT 59604
(800) 821-3874
www.sweetgrassbooks.com.

Printed in China.

17 16 15 14 1 2 3 4 5 6 7

Foreword

George Rohde considered this book to be his life's work. He was born in 1945 in Fargo, North Dakota, the child of parents whose families had been Dakotans for generations. One of his grandfathers rode with the Marquis de Mores as a young man, and his parents were avid outdoor people. As a boy, George was a hunter, a fisherman, and a watcher. What he watched was the life of the prairie. Having graduated high school in Fargo, he went to the University of North Dakota with the intent of majoring in business, thinking that he would move back to Fargo and to an active and successful family business. The problem was, he found out he didn't like business. His passion was, and would be for his life, the northern prairie, its grasses, plants, flowers, birds, insects, fish, reptiles, and geology, geography, and weather.

He switched his academic focus from business to the study of biology. In those days, George and I were acquaintances and casual friends, but we were both biology majors and when we got together we often talked about biological systems. Whereas my biology was biophysics and molecular genetics, his was early and late season grasses, salamander migrations, fresh water crustaceans, and the lifestyles and tribulations of living things. He liked things he could touch and smell and see and watch. He was a field biologist. When he graduated from college, he spent time at the Northern Prairie Wildlife Research Center in central North Dakota and immersed himself in the study of waterfowl, their reproduction, and their predators. Then, and for the rest of his life, he was fascinated by the struggles for survival of the prairie species.

At this time in his life, he encountered cancer and experienced his own survival struggle. He persevered, and won out after a long course of treatment, although the long term consequences of that treatment would slowly limit his physical strength some forty years later. After his reckoning with mortality, he decided that he would spend his life doing something he loved to do every day, and that something was being out on the prairie and observing its life playing out. A salient question was, however, how was he going to make a living?

He thought through the possibilities of farming, ranching, law enforcement, and finally settled on photography. He had always been a perceptive and skilled amateur photographer. After all, photography is really a part of watching, and George was one of the world's best observers. He got serious about photography, and although he was self-taught, he learned through trial and re-trial, error and re-error the way most artists do. He stayed with photography for his lifetime, but it was his medium of expression, not the point of his life. He did not photograph weddings, nor family portraits, nor graduations. Photography was the means by which he sought to portray the struggles for life and the ebb and flow of the plants and animals in the northern plains. To George, the setting of a photograph and the idea that it portrayed real life and real behavior was as important as the image that was produced. He never recorded a staged situation engendered by bait, threat, or some captive animal. The point of every photograph was a recording of natural behavior, not altering behavior to achieve an image.

Although photography was his primary means of expression, it was not the only one. He was a skilled creator of dry flies, truly an artistic pursuit. He made blue-wing olives using feathers from a sandhill crane, mayflies from the wings of a sharp-tailed grouse, and streamers and nymphs from jackrabbit fur. He was also a cook, and the focus of his culinary art was foods like wild plums, wild onions and garlic, wild rice or morels from Minnesota, and the wild game he so respected.

Photography of the prairie is not a way to become wealthy, but it was that profession that allowed him the freedom that he so valued. To market his art, he commonly went to craft shows where he and other artists would direct market the products of their efforts to the public. Early in his career he realized that living in the northern prairie is a different economic environment than, say, the Silicon Valley, and he needed to be able to produce his art in an affordable form. So, throughout much of his photographic career he produced many of his images in the form of notecards that were very accessible, very affordable, and very special. His goal was to allow these images to be part of people's daily lives, and when he began to conceive of this book, he meant it to be based on the images on those notecards and the stories they portrayed.

Another part of the genesis of this work was his relationship with Max, his grandson. Max and George were, in many ways, kindred spirits.

When Max was young, and at that age when the world opens up to the questioning mind, George spent many hours answering questions about the natural world and the creatures in it. Although the book is written in a language a child can understand, this is not a book just for children, but rather one for all those who see, wonder, and imagine.

The extremes of the seasons in the northern plains are dramatic, and George organized his book around those changes in the seasons and the ability of the birds, plants, and animals to survive. The text is reflective of the thoughts that went through George's mind as he watched the evolution of the year, and as he watched the plants and animals change with it and win or lose their struggles for survival. In adulthood, it became my privilege to spend many hours with George sometimes hunting or fishing, but most commonly just watching—and to listen to his observations and his questions of what he thought might be going on in an animal's mind. Now there are those who think "not much," that animals do not have complex thoughts. George believed that on some level that they did. Questions like "what does he think he is doing" when watching a salamander struggle over rocks, or a jackrabbit sitting in the snow in January in a blizzard, or "just imagine trying to find a handful of seeds" watching a Hungarian Partridge in a snow-covered field of wheat stubble, were part of the wonderment that led to this work.

As the long-term complications of his cancer treatment began to catch up with him, George devoted more and more time to the gathering of his images of forty years into this book. Unfortunately, the book was not quite finished before George's allotted time ran out. The text for December had not been completed, there were a few images that had not yet been included in a few of his chapters, and some minor editing was needed. Many of these tasks were completed by Dr. Erik Fritzell, George's lifelong friend and a colleague during their days of researching the prairie, and George's wife, Beverly. They have captured George's voice and intent in a way no others could have done.

Nonetheless, the book is really all George, a compilation of images, observations, and thoughts of a kind, gentle, and perceptive man who spent his entire life watching and wondering about the Northern Prairie.

Bruce Lytle
Cleveland, Ohio
August 7, 2013

PREFACE

George Rohde spent much of his life observing and admiring the flora and fauna of the Northern Prairie. And he loved to share his enthusiasm and curiosity about natural history with anyone that seemed to have an interest. He believed that if others would just take the time to observe the natural world about them, they too would love the prairie landscape and its denizens as he did. For decades, his photographs provided a means to share interesting stories. Periodically he would make a 3×5 bi-fold notecard from one of his images, write a short note about the subject or the "back story," and send it off to anyone who might care. Those notecards and the thought behind them provide the foundation of this book.

When George Rohde died in May 2011, he had nearly completed this book. He had organized selections of his photos into "months," many of which were laid out in notebooks. Most images were accompanied by text designed to help the reader interpret the image and learn about the Northern Prairie and its inhabitants. Based on his notes and several earlier conversations, his intentions seemed clear. Furthermore George's "notebooks" of material were sufficient to allow his sister-in-law, Barb Solberg, to compile a "draft" hard cover version of *A Prairie Year* as a birthday gift for George's wife, Bev.

Unfortunately, George's organizational scheme for his original high quality images was as haphazard as some other elements of his life. It took weeks of searching through piles of discs to locate quality images for each of his selections for the book. Some, for which hard copies were extant, could never be located.

I have tried to complete *A Prairie Year* in a fashion that is faithful to George's original intent. All the images have been selected by him—to one degree or another—for inclusion into the book. Most of the text was originally crafted by George and my editing was mainly to reduce length and enhance clarity. Several "months" lacked a full array of photos and some "months" were missing portions of text. December, for example, consisted only of preliminary lists of possible images. For that chapter, I selected the final group of photos, wrote a short introduction, and provided names of the subjects or a short caption for each image.

I made one "addition" that included text not specifically written for this book, although penned by George. The tiger salamander image (see April) was not among the preliminary images that George was using for the book, but it was included here because it graced a notecard that he sent to me. The accompanying text is authentically George's.

Erik Fritzell
Grand Forks, North Dakota
August 14, 2013

INTRODUCTION

For many years I have been observing and photographing the animals and plants that make up the vast, dynamic ecosystem of the Northern Prairie. Each year may be dramatically different from the preceding. A drought may follow normal precipitation. A winter of little snow may follow a winter of severe cold and much snow. A season may start dry and finish with above normal moisture. Populations of the inhabitants may also vary depending on conditions. Because of the variable conditions, the inhabitants have adapted to change and are able to sustain viable communities.

The entire prairie ecosystem, encompassing the middle third of North America, is comprised of three general types based on the height of the dominant plants, the grasses. From east to west a traveler would pass through the Tall Grass, the Mixed Grass or Mid Grass and finally reach the Short Grass Prairie. The average annual rainfall declines from east to west restricting plant growth. I have concentrated my work on the Northern Prairie which stretches from the Prairie Provinces of Canada south to northern Nebraska then east to Illinois and west to Wyoming. Many species inhabit the entire prairie; however, I believe weather and seasonal difference also divide the ecosystem into three north-south regions. The Northern region experiences the most dramatic and definable seasons. Temperatures may vary from 100 plus degrees during summer to bone chilling 35 below zero during winter. The difference between the record highs and lows during February may exceed 100 degrees.

The prairie landscape today is dominated by agriculture. Much of the original prairie has been altered through farming and ranching practices. Human disturbance and the introduction of non-native plants and animals, either intentionally or by accident, are equally characteristic of the region as are the native species I would prefer to photograph. Ring-necked Pheasants and Hungarian Partridges were introduced years ago and today are thriving throughout the region. Many non-native plants are well-established with some being compatible and others being very detrimental to native plants.

Hopefully, a growing interest in identifying, maintaining, and restoring native prairie areas will help provide habitat for native flora and fauna for future generations to appreciate.

When our grandson was five, I thought it would be a good idea to send him one or two note cards weekly showing images of nature found on the Northern Prairie. I took photographs each week and chose those that provided a glimpse into the seasonal activities of the prairie inhabitants. A short text was written to help explain the suspects' behavior. My thought was the project would be a nice way to introduce Max to the prairie and to provide Max and his parents a seasonal account of the workings of the Northern Prairie ecosystem.

I have also sent notecards to other relatives and friends and, with their encouragement, have included some of those images and writings in this book. I am presenting the chapters as the twelve months, recognizing that the year is actually a continuum of change with the start of a month being quite different from the end. My hope is that the project will provide a general survey of the vast diversity of life, climate conditions, and survival strategies throughout a prairie year.

Throughout my life I have taken the opportunity to observe nature. Friends and I have spent much time discussing behaviors we have seen. Often the topic is the question of animal intelligence. We have all noticed a cat sleeping in a sunny window, or a dog looking at the backyard birds. We question—does the cat feel a sense of well-being? Does the dog somehow enjoy the activity? While we cannot say they do, I can say, I hope so. I began to wonder about an animal's sense of wellbeing many years ago while I was studying Red Foxes. On a May afternoon I was set up observing an active fox den. They were unaware or uncaring of my presence.

Foxes remain active throughout the year and must deal with the weather elements. The winter had been very cold with much snow and very little moderation. The April weather was normal and by early May spring's activities were well underway. Towards evening, the female fox returned to the den area and nursed her five pups near one of the den entrances. Shortly after the feeding, she retreated to the edge of the den area and sat on her haunches and looked out over the landscape.

As I watched, she watched a pair of Mallards fly by. The evening was memorable. I have since thought about that moment and the fox and her family. She and her mate had successfully survived a very difficult winter. The snow had been very deep, and one observer had noted that foxes had been eating sunflower seeds as the snow had made it difficult to find mice. The fox not only survived but carried her unborn litter and successfully delivered the five healthy pups that were playing nearby. The food supply was now plentiful as the Meadow Mouse population had surged under the snow pack and the mice were easy to catch. It was a good time to be a fox. She appeared to be relaxed and at ease. I wondered if she had a sense of wellbeing. I hope so.

—George Rohde

A Prairie Year

JANUARY

January is the most dreaded month of the long cold winter on the Northern Prairie. For humans, the attention to the holiday season's activities is past, school is back in session, and the coldest month of the year has started. On many days the weather is windy and brutally cold. We say to people, "Today is a good day to hunker down." We mean it, often for one's safety. Certainly, there are many positive activities and events to enjoy, but often the weather is not one of them.

Wildlife of the Northern Prairie has adapted to the winter conditions with a wide variety of strategies. The target outcome for all species is to survive the winter period and emerge in adequate health to raise the next generation. Seemingly, the easiest method is just to leave the area and head to warmer climes. Other species hibernate or become inactive away from the harsh elements. However, a collection of birds, mammals, and a few insects and spiders remain active throughout the winter months.

The depth of the snow pack or cover can benefit some but cause problems for others. As long as the snow does not melt, it acts as a very good insulator for birds or mammals that burrow into it. In addition, there is a great deal of winter activity beneath the snow and above the soil surface. It is called the subnivean layer. The space may be only a few inches high but may provide a living space for mice and voles and a variety of insects and spiders.

While I do not look forward to the beginning of January, there is a significant turning point during the month. The average low temperature for the year occurs sometime during the third week. The winter starts to wane, slowly perhaps, but with the welcome arrival of the new seed catalogs, spring is out there.

On a very cold blustery January day we may decide that instead of going to a movie or out for supper, we may say, "Tonight let's just stay home and 'hunker down.'" As cozy as that may sound, I think these White-tailed Jackrabbits have a quite different perspective on the need to "hunker down."

Below: On a bright sunny winter day, a Long-tailed Weasel was searching for a meal. Weasels remain active during the winter but spend much of the time in the subnivean layer under the snow pack. They shed their fur twice a year, and where there is a snow cover, its winter pelage is white. Weasels provide a wonderful example of protective coloration. Their white fur coat blends with the snow making them very difficult for predators to see. Also, when the weasels are hunting, their prey may fail to recognize the danger.

Above: The life style of the White-tailed Jackrabbit is a bit of a marvel to me. It is apparent that they rely on their eyesight to detect danger and their speed to escape. Therefore, resting time is in shallow forms without sight obstruction but with little protection from blustery winds or snow. They are adapted to spending their lives exposed to the elements of the weather.

Opposite page: Apparently, their fur provides great insulation as it is not uncommon to see them away from shelter on very nasty days.

Opposite page: Red Foxes are well suited for a winter on the Northern Prairie. They grow a beautiful thick winter coat of fur and have a very bushy tail. When sleeping, they stay warm by curling up and covering their nose and face with their tail. They eat small rodents and rabbits and may eat fruit and seeds when the snow pack is deep and food is scarce.

Left: Occasionally mice will venture onto the snow surface in search of food. This White-footed Deer Mouse has found a seed from the Stiff Goldenrod.

Below: Winter can be a great time to gather with friends and family. Coveys of Hungarian or Gray Partridge take togetherness to another level. It is their way of surviving very cold temperatures. They huddle together, forming a "rosette" to provide some wind protection and to minimize heat loss. Gray Partridge are native to northern Europe and were introduced onto the Northern Prairie. The birds are well adapted to the climate and populations are now established.

Above left: It may seem odd to consider Black-capped Chickadees as prairie residents, but these adaptable birds are common along wooded streams, rivers, cities, and wooded farmsteads. They feed on seeds and readily venture into fields to find sunflowers. They will also feed on meat and fat from a deer carcass.

Above right: Cold winter temperatures present challenges for insect-eating birds. Most insects are not active but have various strategies to survive the winter. The larva of the Goldenrod Stem Fly causes the plant to form a bulb or gall, inside of which it overwinters. Downy Woodpeckers venture onto the prairie in search of the Goldenrod patches. They drill into the galls and eat the larvae.

Left: Several species of birds migrate to the Northern Prairie to spend the winter. Common Redpolls are one of these visitors. They are small birds that are usually seen in flocks feeding on seeds from prairie plants.

Right: The key to surviving the rigors of winter is to conserve energy. Inhabitants spend most of their time resting or sleeping in sheltered places.

Below: A small population of Muskrats is living in the marsh. They have piled bulrushes and cattails into a conical-shaped house which they enter from under water. Although it is not my idea of cozy, Muskrats may be quite comfortable throughout the winter. They have ample food. An insulation blanket of snow will keep the inside temperatures less extreme. As long as a Mink does not find them, the Muskrats will be safe.

FEBRUARY

Winter continues into the second month of the year. The snow depth generally increases and the region may experience periods of very cold temperatures. We need a break, a moderation of the weather. Our hope starts on February 2, Groundhog Day. Some time ago, as legend has it, a woodchuck emerged from its den expecting spring, saw its shadow, shivered and retreated into the Earth's comfort to sleep for six more weeks. It seems as though we need a less pessimistic envoy, something that foreshadows the positive weather change we are anxious to enjoy.

Above: On the Northern Prairie the harbinger season's change is the Horned Lark. A February country drive may be highlighted with the sighting of a small flock of larks along the roadway. These birds have migrated back north from their southern wintering grounds. The weather and scene may feel and appear as mid-winter; however, these small birds are declaring that spring is coming and they are anxious to begin a new season. I think the larks view their life glass as half full.

A herd of White-tailed Deer was feeding in a soybean field. Beans have become a common agricultural crop on the prairie because new plant varieties have proved successful in areas once thought too dry or where soils were too thin to promote good growth. After harvest there is very little standing vegetation remaining, but, there are lots of beans scattered across the field. Since the fields are nearly bare soil, much of the snow is blown off large areas making it easier for deer, pheasants, and grouse to find the beans. The tops of the hills have the least amount of snow, and the many groups of deer are feeding on the beans.

Above: Bison are large grazing animals that require up to 30 pounds of plant material, mainly grasses, every day. They use their beards very efficiently to sweep snow off the food. In deep snow they are able to clear away several feet of snow by moving their massive heads from side to side.

Left: The American Bison or buffalo is a wonderful example of adaptation to the sometimes very harsh prairie ecosystem. The animals grow very thick coats of hair and can withstand brutal winds and low temperatures.

Above: A small flock of Ring-necked Pheasants was scratching for soybeans. The countryside had received significant snowfall and food was hard to find. It is difficult to imagine the effort required to locate a handful of beans that is scattered under six or more inches of snow. If there is a layer of ice or the snow is deeper, the task is more difficult and more energy is used.

Right: Where to begin looking?

Left: Ring-necked Pheasants were brought from Asia and introduced to the prairie early 1900's. They have become very well established in farming areas. The males are very colorful while hens are dressed in shades of brown. The hens' coloration helps them hide from predators during nesting and rearing of the young. Pheasants can survive cold winters quite well; however, they need shelter from the strong winds and access to seeds and grain.

Opposite page: Pheasants often escape by running, but they don't escape looking a little silly.

Below: Eagles and hawks begin to drift northward during late February. It seems a bit unfair that a Pheasant would struggle through a harsh winter only to end up as a meal for a returning Bald Eagle.

March

During the first days of March, temperatures begin to warm and usually the snow pack starts to melt. Our thoughts are in high gear for the coming spring. March Madness not only means basketball tournaments, but it also means we are tired of being cooped up inside and are anxious to enjoy warm temperatures and outdoor activities.

The saying is that March weather "may come in as a lion and go out as a lamb." However, there seems to be no normal for March. It is a month of struggle between winter and spring. Some years there may be little or no melt of the snowpack and the bird migration may have to wait. Other years the temperatures may rise, and by the end of the month it may appear as though spring has arrived. Occasionally, the region may receive a setback of a violent snowstorm.

Even with the uncertainty of the weather, mammals and some migrant birds seem anxious to be finished with winter and to start the activities of spring. A few individuals may migrate into the region or emerge from their burrows. They may be rewarded with a run of nice weather.

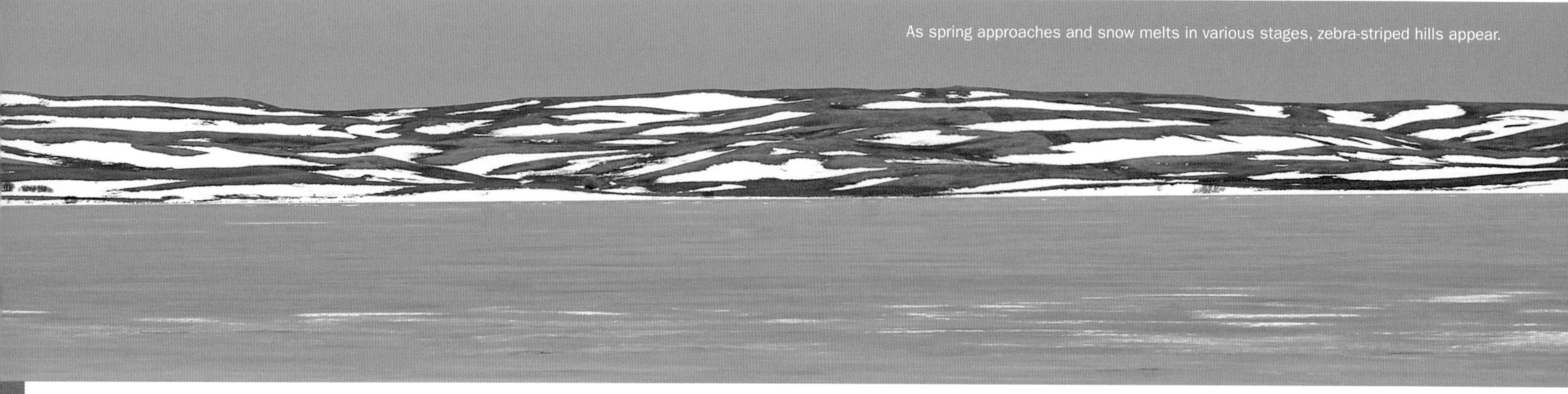

As spring approaches and snow melts in various stages, zebra-striped hills appear.

Early in March the Horned Larks begin their courtship activities. A male lark was singing his song from a wind-sculptured snow drift. The landscape appeared to be mid-winter; the lark knew differently. The genetic songwriter has crafted the notes of the lark to inform his suitor that this area will be a wonderful place to raise a family. The notes tell the story that within a month the snow will be gone and temperatures will warm. More importantly, there will be nesting sites, seeds for the parents, and insects for their young. It is a complex message. Horned Larks are the most abundant nesting birds in the region.

Above: As the month progresses, more and more residents and migrants make appearances in the region. This Raccoon was headed to a small patch of water at the edge of a marsh. I would imagine it was very hungry after spending the winter sleeping under the snow pack or in an old building.

Left: Mink spend most of the severe winter months beneath the snow pack. However, they remain active and can be seen along streams where the water does not freeze. This female stopped for a drink from the flowing river. Minks and weasels are closely related, but minks keep their brown fur color throughout the year.

Above: Patches of snow remain, but on a warm sunny spring afternoon this Richardson's Ground Squirrel has emerged and is eating last summer's plants.

Above right: Towards late March, not only are more animals active, but their purpose begins to include preparations for the birth of their young. The Black-tailed Prairie Dog was busy gathering bedding material to comfort its soon-to-be-born pups. Prairie Dogs dig complex burrows or tunnels that may be fifteen feet deep and extend laterally for one hundred feet. They live in colonies or towns with many individuals sharing the duties of watching for predators.

Opposite page: As the effect of the March sun continues the snow melt, ground squirrels wake from hibernation and emerge from their burrows for a glimpse of spring. Even beneath snow drifts they somehow sense a change in the seasons. This Thirteen-lined Ground Squirrel tunneled through three feet of snow for a look and some fresh air.

About a dozen or so different hawks migrate across the prairie during spring and fall. Some nest in the region while others head farther north. They feed on mostly rodents such as mice and ground squirrels. Their migration pattern seems relaxed as they drift on air currents and survey the countryside. This Rough-legged Hawk is pushing north as the snow pack retreats.

Early spring may be a difficult time to find food for some returning bird species. It may require scavenging fish and mammals that had perished during the winter. The Ring-billed Gull had found a White Sucker that had been preserved in the ice covered pond. *Yum!*

April

Winter is a long and very difficult period. Even today, with advances in clothing, home heating, and transportation, the Northern Prairie winter can be an uncomfortable and dangerous time. I can only begin to imagine what life was like for the Plains Indians and early settlers. By the end of March, food and fuel supplies were probably exhausted. Days would be getting longer and temperatures warmer, but a brighter future was dependent upon the coming growing season. Plant growth is the key to the success of the prairie inhabitants. How uplifting and powerful the sight of the Pasque Flower blossoms must have been.

I would imagine a positive mood change swept through many communities when a small bouquet of crocuses was presented. How many small fingers picked bunches and proudly gave them to a teacher in a rural schoolhouse? How many giggles were heard as children in moccasins used the fuzzy blossoms to tickle the noses of their siblings? How many times did callused hands gently gather some blooms, place them into a prized glass, and set them on a crude dinner table? The mood swing must have been palpable. A new season has begun and with it hopes for good weather and bountiful harvest. The future looks bright and life will be better.

Left: Pasque Flowers are blooming! These "Prairie Crocuses" are proof that the soil has warmed and the growing season has begun. By flowering early, they escape the hot dry conditions common throughout the region.

Opposite page: While only 3-5 inches in height, Pasque Flowers stand tall over the grazed pasture grass surrounding their stems. I think it is probable that the ancestors of today's flowers were rooted in the very same soil. Their cells may contain nutrients cycled through buffalo that grazed the hilltops centuries ago.

Even a chilly wind provides a lift for a Northern Harrier.

Above: A Trumpeter Swan pair can be very aggressive when another swan gets too close. The swan looks like, "What did I do?" I know the feeling. I grew up with two older sisters, and I remember often being the target of their mischief. Trumpeters disappeared from the prairie early in the 20th century but have been successfully reintroduced into the region. It is a treat to see these large birds nesting, where they belong, on the Northern Prairie.

Left: For a Muskrat, having lunch in the sun by water's edge is a rite of spring.

Right: Male Red-winged Blackbirds arrive on the northern prairies well before the females do. They proclaim to others "this is my territory, and it is grand."

Male Greater Prairie Chickens dance to their own drumbeats and display their finest feathery forms to impress the females. Some Native Americans copied them in their own dances. And what about the "Funky chicken?"

Right: Subtle, yet elegant, the Horned Grebe.

What's not to smile about—It's Spring! Tiger Salamander.

For Northern Pintails, twisting, turning courtship flights fill the sky.

Snow Geese are busy heading to the far north before beginning to nest.

MAY

Serious breeding activity dominates the Northern Prairie during May. The landscape will have "greened up" by the end of the month, and food and cover will no longer be sparse. Perhaps more than any other month, the richness of the prairie becomes evident.

Left: An American Bittern male is making a mating call. Bitterns live secluded lives among the cattails in marshes. Often they are heard and not seen. When they detect potential danger, they extend their head and neck upward. Their feather pattern and behavior result in an effective camouflage strategy. Yet they keep their eyes on the danger.

Opposite page: Does the emergence of flowers have significance to any of the wild inhabitants of the prairie? Bobolinks certainly would be aware. Does it mean anything to them? Do they stop and look? I hope in some way the wild beings enjoy the moment.

Above: White Pelicans use several strategies to obtain food. They eat fish and amphibians which they find in lakes and rivers. Occasionally they watch Double-crested Cormorants, and when one catches a fish, the larger pelicans chase the cormorants. The pelicans are much larger and very aggressive. Sometimes the bird drops the fish and a pelican gets it. But this cormorant seems to have a better strategy. That is, to eat and flee.

Left: During the last decade, the Great Egret has been extending its breeding range onto the Northern Prairies. The area has experienced a dramatic increase in precipitation resulting in more lakes that support fish and amphibian populations. Egrets have taken advantage and are now found nesting in these habitats. The birds construct their large nests with sticks carried one at a time.

Right: When migrant birds return to the marshes during May, the activity and sounds increase. The most energetic and noisy are Marsh Wrens. The males arrive first, establish territories, and proclaim them with song. They may be difficult to see, but their singing is welcome news that they have returned. Males also construct many nests throughout their territory. This male divided its time between singing and nest construction. The dummy nests are poorly constructed and their purpose is a mystery. One idea suggests that it may be a means of expending energy. Nest building stops when the females arrive.

Opposite page: Prescribed burning is often used to manage a healthy prairie ecosystem. Native plants are well-adapted to surviving fire. The dried vegetation burns off, but the viable underground portions of the plants are protected by the soil. Usually fire is used in early spring to remove accumulated plant material and to kill unwanted grasses like Kentucky Blue Grass. Although scorched prairie may seem devoid of life, it is anything but that. Within a week or so, new plant shoots can be seen. The burned area will soon be covered with wild flowers and warm season grasses such as Big and Little Bluestem, Switchgrass, and Indian Grass. After a fire, it is a concern as to what happened to the animals. This Thirteen-lined Ground Squirrel seemed to have managed just fine. It must have retreated to its burrow and escaped the flames. Now it has a new crop of grass to eat and a place to raise its young.

Above: A buffalo calf seems hungry and is trying to pressure Mom for some milk. The little calf gently nudged its mother and even rubbed heads. The behavior was fun to watch as it clearly resembled any child pleading with Mom for a favor or treat. Eventually and with a look of "okay, enough," the cow stood up.

Left: The little buffalo seemed content for it had "Got Milk." Bison calves are born during late April and early May. They are dark brown when born and soon the hair is replaced with a rusty coat.

Opposite page: A buffalo bull grazed on the new grass shoots of spring, while Brown-headed Cowbirds grabbed insects as the big bull forced the bugs to move out of his way. Historically, cowbirds followed the massive herds of buffalo as they fed over the prairie. Perhaps the nomadic lifestyle of the birds led to their brood parasitic behavior of laying eggs in other birds' nests and relying on them to raise the young.

Above left: A Yellow Warbler used a Phragmites stalk as a perch while it watched for midges to surface on a nearby pond. It would then fly down and grab a snack.

Above right: Migrating warblers of many kinds can be seen feeding on gnats along the windy shore of lakes and ponds. The insects come out of the water and the wind blows them to shore where they accumulate. This Yellowthroat has found a meal of gnats caught in a spider's web.

The days continue to warm as May progresses. Ice on the lakes has melted and as the water warms, insect life becomes active. Gnats are a non-biting, mosquito look-a-like that spend most of their lives in the water. They emerge as flying insects to mate and lay eggs. About mid-May, there is a huge hatch and flight. Migrating birds, such as this Tree Swallow, take advantage and can be seen feasting on the emerging gnats before the insects are able to fly.

Many species of birds, while crossing the Northern Prairie, interrupt their spring migration and stop to rest and to feed. There can be wonderful opportunities to see birds that otherwise are not found in the area. These Ruddy Turnstones have found a ready supply of gnats along a large prairie lake. The birds probably wintered along the Gulf Coast and were headed north to nest on the tundra.

One of the rarest of the prairie summer residents is the Piping Plover. A recent estimate is that there may be only 3,000 nesting pairs in all of North America. There are two populations of the sparrow-sized birds. One group nests along the East Coast and another chooses to nest on the Northern Prairie. A few of the prairie birds can be seen on the gravel bars near alkali lakes or found on sand bars of the Missouri River. Males arrive first and establish or claim a territory.

Above left: Marsh Wren females arrive a week or so after the males. They choose a mate and begin their summer chores. Nests are built among and attached to cattail stalks or other marsh plants. They are made of several layers of vegetation and are lined with feathers, or cattail "down." Marsh Wrens may raise two broods during the summer.

Above center: While birds are singing on the cattails above the wetland, toads are at the water surface using the same strategy to attract a mate. This Canadian Toad was one of several males courting females. Thousands of eggs are laid by each female. The tadpoles spend about six weeks in the water then emerge as small toads.

Above right: When May flowers appear, Bumblebee queens are sure to follow. This queen is landing on a Shell Leaf Penstemon. The bees have emerged from hibernation and will soon begin to establish a hive, usually located underground. The queens are the only survivors from last summer. They soon lay eggs and feed the larva nectar and pollen. The first young are workers about half the size of the queens. Eventually, the next generation of queens and males will be raised.

Opposite page: Last fall Monarch Butterflies left the prairie and migrated south to overwinter in Central Mexico. They started north in February and March reaching Texas and Oklahoma as wild milkweed plants had begun to grow. Adult Monarchs bred, laid eggs, and the young caterpillars fed on milkweed. The next generation was underway. As the spring moved north, the new Monarch brood followed. During May, they arrived on the Northern Prairie, sometimes appearing worn and tattered from the flight. This female is feeding on nectar from a Wild Parsnip flower. Soon she will mate, lay eggs, and her caterpillars will feed on the growing milkweed. Her task is then complete.

June

Major additions to many animal populations dominate prairie life in June. Litters are born and broods hatch. By month's end young of the year will appear for the human observer. One only need to go out on a calm evening to experience the importance of insects to prairie ecology—oh the mosquitoes!

Above: Monarch Butterfly eggs hatch after four days and the young caterpillars begin to feed on milkweed plants. The small caterpillar is only a few days old but is growing rapidly. In about two weeks it will form a chrysalis or pupa stage. After two more weeks it will emerge as an adult butterfly.

For newly born White-tailed Deer, survival may mean “freeze,” and look like a pile of leaves.

Left to right:

Nesting activities are well underway for most birds, including Yellow Headed Blackbirds. Females do the nest building, and this one has gathered pieces of cattail stems. She will weave the nest among cattail stems that have grown above the water.

This little Killdeer is only a few days old. It is hard to imagine that all that body and those long legs were packed into an egg. The young are protected by the parents, but the chicks are able to recognize and catch insects for food.

Often during the summer, sunrise is a calm and very pleasant experience. This Eastern Kingbird seemed content to let the sun add its warmth while the day began. The bird was in no hurry to start the frenzy of catching insects for its young. After all, the insects were also letting the sun warm them.

Above right: Cliff Swallows build their nests with mud they gather from water's edge. Nests are built in colonies usually under the shelter of a cliff or bridge. Several species of swallows nest in the region all with different nesting strategies. Some choose to nest alone while others prefer groups. All of them feed on insects, often in mixed flocks.

Below right: An American Coot has led its brood to the marsh shoreline to search for food. Coots eat both plant and animal matter. The young are not thought of as the cutest of the marsh babies, but they do have their place and are successful nesters on the Northern Prairie.

Opposite page: These two young Thirteen-lined Ground Squirrels were wrestling near their burrow entrance. It was very near the burrow of the ground squirrel photographed on the burn a month earlier. Perhaps they are its offspring. The young on the right was angry, or was excited about its first tooth . . . maybe both.

Left: The Little White Ladyslipper is a beautiful small wild orchid found on moist Tall Grass Prairie. The flower is a cousin of the larger Ladyslippers of forest habitats. The Little Whites are perennial and may grow in bunches or single plants. They flower during June.

Opposite page: Sometime during June, Western Grebes chicks begin hatching, leave the floating nest, and ride on the backs of the parents. The adults' wings provide protection for the young from the weather elements. Both parents care for the chicks, taking turns between hosting and finding minnows to feed the hungry young.

Above: On a late June afternoon, the wind has calmed and hen ducks have escorted their broods from cover and out to the open water to feed on invertebrates. Two years before, the wetland and others like it, were dry. Very few, if any, duck broods were raised in the area. However, winter snows were deep and spring rains helped refill the wetlands. Ducks responded with a strong nesting effort. It was a thrill to see the response to the wet cycle.

Right: A brood of Trumpeter Swans is looking for food with the help of a parent. The adult locates a likely feeding spot. Then by fluttering its wings and paddling its feet, aquatic invertebrates are exposed to the watching young cygnets.

Far right: There had been very beneficial rains and the prairie grasses have had a productive growing season. A sure sign is the need for this Richardson's Ground Squirrel to climb on a rock in order to look over the prairie.

JULY

By the end of July, the prairie landscape begins to turn brown. Water levels in the abundant marshes decline creating new environments for aquatic plants and invertebrates. Spells of hot weather provide new challenges for young just learning to cope with life itself.

Above: How many of us have experienced some setback in life? This Purple Coneflower suffered a severe injury while growing. Somehow its growth form compacted, remained in balance, and the plant successfully flowered and produced seeds. It is a survivor.

The second generation of Monarchs has emerged during early July. The butterflies feed on the nectar from the flowers of milkweed, Gayfeather, thistle, and Blazing Star. Soon they will breed and lay eggs; then the third brood will hatch.

Above: The cool season grasses have formed seed heads and these have ripened. A Thirteen-lined Ground Squirrel was busy gathering the seeds. Its cheeks are bulging with grain. Grasses and flowers ripen throughout the summer, and several rodents tkeep busy storing food as it matures.

Right: A male Sedge Wren can be seen singing from a willow branch. Many other birds have ceased their singing when we get to the days of July. However, the little wren may have recently arrived on the nesting grounds, and its breeding season is just starting. These birds have a curious strategy of nesting in an area one year and may move to another the next year. They prefer moist tall grass areas and avoid cattail marshes.

Opposite page: Gayfeather

Right: The adults emerge during early summer and stay active into September. The adults feed on the nectar of many kinds of flowers including milkweeds and thistles.

Below: Regal Fritillaries are mid-sized butterflies found only on Tall Grass Prairie. There is one generation during the summer. The eggs hatch during the fall and tiny caterpillars spend the winter in the ground leaf litter. The strategy of the caterpillars to overwinter enables them to feed on violets, which are early season plants. Regal Fritillary caterpillars will eat only prairie violets, and since the plants are found only on the Tall Grass Prairie, their range is restricted.

An American Wigeon hen watches her brood as they search for food. While not common, these medium-sized ducks nest on the Dakota prairie. The adults sometimes feed with Tundra Swans and steal food that the swans have brought to the pond surface.

Pronghorns inhabit the mixed grass and short grass prairies. They are beautiful mammals with great eyesight, and they are very fast runners. They use their speed to outrun predators. This small group consists of a female, or doe, with her two young. A male, or buck, has joined the three to feed on the grass shoots.

Above: Ferruginous Hawks are found nesting on the arid to semi-arid Northern Prairies. The hawks may build their nests on the ground because much their range is treeless. Historically they used buffalo bones to form the outer nest. These four young are about to learn to fly.

Opposite page: The Plains Pocket Gopher is a common mammal, but one that is seldom seen. It has a fossorial lifestyle, meaning it lives underground. In fact, these gophers spend their entire lives in tunnels only to emerge during the spring breeding season and occasionally to find food near the tunnel entrance. It is interesting to see how clean and groomed the fur was considering where and how it lives.

Above: This Western Grebe has a problem: one fish and four mouths. By mid-July, young grebes are swimming and diving but still very dependent on their parents for food.

Left: A female American Goldfinch has collected cattail down for the inside lining of its nest. Goldfinches are late season nesters—usually starting nesting activities in July when their food supply is most plentiful. They feed the young the seeds from sunflowers and thistles, and occasionally insects.

Opposite page: For the Great Blue Heron a tasty morsel, for the Leopard Frog a predicament, at best.

August

The activities from late August and into November are concerned with preparations for the coming winter. Most of the birds and even a few insects migrate through and away from the Northern Prairie. Many of the mammals and all of the reptiles and amphibians retreat into the earth or wetlands to find shelter from the soon-to-be freezing temperatures. It is interesting to study the strategies used by the diverse life forms to avoid or overcome the rigors of winter. If these animals are to successfully reproduce next spring, they must not only survive the winter, they must return in good health and to habitat that is diverse and abundant.

Right: Monarch and Blazing Star

Opposite page: A Pied-billed Grebe juvenile has found a Giant Water Bug for its lunch. The grebe is on its own and quite capable of finding food. These grebes usually inhabit smaller wetlands and ponds than do their much larger cousins, the Western Grebes. Pied-bills are able to catch and consume insects and amphibians that are larger than I would expect.

Above: If you find a stand of wild sunflowers, you will likely find American Goldfinches feeding on the seeds. The males give us a nice example of protective coloration. The yellow and black of its feathers match with the colors of the flowers.

Left: A Solitary Sandpiper has found a small pond in a meadow where it can find food. These sandpipers are one of the first migrants to cross the prairie on their southern flight. The birds nest north of the prairie and winter in Central and South America. They are the only species of North American shorebirds to nest in trees rather than on the ground.

Above: A young Sora is feeding along the edge of a wetland. As with most rails, this species is very secretive and rarely ventures away from protective cover.

Right: Pronghorn males have a scent gland located in a black cheek patch behind their jaws. The gland produces a musky odor. This buck is rubbing the cheek patch on a thistle flower to leave scent and thus marking its territory. Mating takes place during September; however, bucks become active during August and start to gather a harem of does.

Opposite page: On a hot August day on a dusty gravel road somewhere on the western Short Grass Prairie, three friends were treated to a rare encounter with a Swift Fox. These small foxes are seldom seen as they are mostly nocturnal and their population numbers are low. As we were to learn, the females are territorial, which is unique among Canids. I think the question of whether or not a wild animal can communicate with us was answered. She looked into our eyes and quite clearly told us, “It was nice of you to visit, but this is my turf and it is time for you to move on.” We got the message.

SEPTEMBER

September throughout the Northern Prairie begins a dynamic time for its residents and visitors. There is an obvious shift of activities from summer to fall and winter. Schools are back in session. We have fall sports, and hunting and fall fishing take up our time. In addition, we harvest our gardens, put hoses away, rake leaves, and generally get our homes ready for cold weather and the coming winter. We even make sure the snow blower is tuned. And, of course, there are very significant changes in plant and animal life.

April and May are the months when the Northern Prairie Region fills with life; September and October are the opposite. There is an emptying of the region. Some residents migrate, plants become dormant, and some resident animals hibernate. Many species of birds travel through, stopping to feed and rest, on their way to warmer climes.

Above: One of the largest nesting sites for White Pelicans is on the Chase Lake National Wildlife Refuge located in central North Dakota. Thousands of the large birds return each year to raise their young and feed on fish and salamanders found in the prairie lakes. Three pelicans appear to be observing, with varying degrees of interest, another's fishing technique. Pelicans usually begin to migrate by the end of September.

Opposite page: September is a month highlighted by change. Perhaps the first sign is a shift in the prevailing wind direction. Generally, summer winds are from the south. When the winds switch to the northwest and gain intensity, cool air is ushered into the region, and we say, "It feels like fall." It is.

Above: During the warm days of September, it is possible to observe a feeding White-lined Sphinx moth. The insects resemble hummingbirds as they hover feeding on flower nectar.

Left: Early September brings the start of the Monarch migration. Native Tall Grass Prairie sites are good places for butterflies to forage for nectar on the late-blooming Blazing Star flowers.

Right and below: Bumblebees remain active during September. The males or drones are very busy with attracting the queens. The drones gather in courting groups with each male choosing resting posts on the tips of plants or sticks. There is much activity as they fly off to chase other males then return to the perch. When queens are attracted, mating takes place. Eventually, only the queens find a place to hibernate and survive the winter. The impregnated queens will emerge the following spring, find a nesting spot, lay eggs, and raise a new brood.

Right: On a beautiful mid-September afternoon, a Monarch butterfly drifts southward over eastern Kansas. Perhaps it has come from the Northern Prairie, and it is now slowly making its way to the central highlands of Mexico. There it will winter with millions of other Monarchs. Next spring it will head north and begin the succession of the next generations.

Far right: A Baird's Sandpiper has stopped to feed on a prairie alkali lake shoreline during the sandpipers' fall migration. This species nests in the far North and winters in South America. It is a treat to view such a traveler. Small sandpipers, called "peeps," are very difficult to identify in their fall plumage. It is easier if other species are nearby and can be compared.

Below: Two Blue-winged Teal share a loafing bar on a September afternoon. Waterfowl continue to be in a post-nesting pattern with days spent feeding and resting on their brood marshes. Later in the month they will begin to congregate and migrate.

Above: This male American Goldfinch is feasting on ripening wild sunflower seeds. The bird has started to molt from its breeding plumage into eclipse plumage. The brown feathers on the wing are new. The feathers will be molted again towards spring.

Left: While it is rare to find hummingbirds nesting on the prairie, they do migrate across it. This young Ruby-throated Hummingbird found a patch of Jewelweed and fed on the nectar. The Jewelweed patch is located along a small spring uphill from a marsh. Annually, this area provides a feeding and resting stop for several hummingbirds.

Left: Fall brings large numbers of hawks into the region. They feed on rodents, birds, and amphibians. Thirteen-lined Ground Squirrels begin hibernation and are seen only occasionally, usually on warm sunny days. It makes sense that they keep their heads down as the hawks are hungry and watching.

Far left: A Prairie Falcon looks over its hunting area on a sunny autumn morning. The annual hawk migration is well underway and many species of birds of prey can be seen hunting, and drifting south across the region.

Below: On a sunny calm September afternoon, this female Regal Fritillary was sipping nectar from the wild Aster flowers. Her wings are worn, tattered, and faded, but she has survived the summer and completed her task of mating and laying eggs. Her young caterpillars have hatched and will spend the winter in the leaf litter. This female was at a different stage of life. Gone was the rapid pace of summer. Perhaps she was experiencing a sort of genetic peace.

Sometimes it may take a while for young animals to learn how to find food. They may have to take advantage of unusual resources. The young Red-tailed Hawk discovered it could make a meal of crickets it found on a grazed pasture. I would guess the hawk was having difficulty finding and catching rodents – its usual prey.

Right: Ice crystals have formed on the petals of a Curlycup Gumweed flower. A killing frost has taken place. It is one of the most important events of a prairie year. Generally it happens on a late September clear calm night. The temperature falls to the upper 20's F. The growing season ends.

Below: Franklin's Gulls stage along the shoreline of a large prairie lake. The birds congregate and move south in large flocks, resting and eating as the travel. They eat mainly insects and must move ahead of the frost line.

From mid to late September, Sandhill Cranes move onto the Northern Prairie. The large birds can be found in large numbers near shallow lakes with wide mudflats or exposed shorelines where they roost. They rely on their eyesight to see dangers and thus prefer open areas. During early morning, they fly to grain fields and pasture land to feed on waste grain, grasshoppers, and crickets. Their calls are very melodic and can be heard at some distance.

October

October continues the transition toward winter on the Northern Prairie. During most years, a killing frost has taken place. The majority of the flying insects have disappeared, and the remaining flowers are drying and their seeds ripening. Food is scarce for birds looking for nectar and small insects. Consequently, many species of birds have migrated south.

Microclimates around prairie lakes are warmer than are the uplands. Waterborne insects and other invertebrates are present in the lakes. Small fish continue to be found along the shallows. Migrating shorebirds pass through the region; some stop to feed and to rest.

A clear, calm frosty October morning should be experienced annually. As sunrise approaches, the sounds of the stirring prairie residents are exciting. There are pheasants and sharp-tails from the uplands as well as cranes, geese, ducks, and swans from the marshes and lakes. It is possible to hear wing beats as the geese rise from the lakes and head to the grain fields. The amount of activity is very impressive. The morning also provides pleasure to our other senses. The humidity accentuates the wonderful smells of sage and a collection of other prairie plants. A drift of friends' coffee brings warmth to the morning's chill. An October day is ahead, and we are there to embrace and enjoy it.

Opposite page: The time is right to witness the dynamic changing Northern Prairie. Waterfowl take center stage. Many species of ducks congregate on the lakes. Some feed on waste grain, while others find feed in the marshes. Huge flocks of Snow Geese migrate into the area. These geese spend several weeks feeding on grain and building fat reserves for their move south.

Left: Northern Prairie Water Pipits nest on the tundra and then winter in the southern states. This bird was part of a small flock that spent a couple of weeks feeding on small insects found along the shoreline of a prairie lake.

Opposite page: Shorebirds take advantage of the wind when feeding along a lake's shoreline. Wind and wave action concentrates the food. This Greater Yellow-legs has stopped to feed on invertebrates. The bird nests north of the prairie and is on its migration to the southern coast.

Below: The young Black-crowned Night Heron is a bit creepy. It should be real creepy to any frog in the small wetland. Herons are very good at sneaking quietly along the water's edge in search of minnows, large insects, and amphibians. These birds build floating platforms made of cattails. Their nests are then built on the cattails.

Above: An Avocet has found a tidbit while searching along a prairie lake. This bird is dressed in its fall plumage.

Right: The flock of Avocets has gathered along the windy shoreline of a large prairie lake. They are watching for and finding small invertebrates on which to feed. These large shorebirds nest on the prairie from Canada to Texas. They winter on the southern coasts.

Above left: During October large numbers of geese migrate into the region to feed on the waste grain left in the farm fields after harvest. Avian predators follow the southern push of geese looking for an easy meal. This immature Bald Eagle perched in a tall Cottonwood tree along the edge of a large lake. Snow geese were using the lake as a resting area.

Below left: Swainson's Hawks are common nesting hawks on the prairie. They prey on rodents but will take frogs when necessary—as this one is doing. Frog populations thrive during a wet summer, and by fall they are easy targets for hawks, herons, and egrets.

Opposite page: This Burrowing Owl is probably the last of its family to head south. These small owls are migratory, wintering in the southwestern states. Their diet consists mainly of insects, such as grasshoppers and crickets. Small rodents and birds are also taken.

Above: These Snow Geese have migrated onto the Northern prairie from their nesting grounds in the Arctic. It is a sure sign that the season is late October when these birds arrive in large numbers. Sometimes thousands of geese can feed in a field, and many thousands may roost on a nearby lake. "Snows" and "Blues" are color phases of the same species. A lone White-fronted Goose is feeding with the other geese. Can you find it?

Opposite page: It seems as though there is a standoff between a Ring-necked Pheasant and a Coyote. I wonder if both are young of the year and a bit inexperienced. Or, perhaps they are both enjoying a beautiful October morning and not interested in disturbing the peace. After a short time, the Coyote jumped down and walked off. The Pheasant, maybe a little naive, walked over and looked over the edge. They may meet another day.

Above: Canada Geese feed on grasses during the summer. In the fall they switch to waste grain. New varieties of soybeans have enabled farmers to plant where previously beans would not have produced a good crop. Geese, ducks, and upland game birds have discovered these crops and now are commonly seen feeding in the fields.

Left: October is the month when ducks and geese are on the move. Some migrate south while others congregate on larger bodies of water to feed and prepare to move south. The birds, like these Canvasbacks, are very active at dawn and dusk. They fly from their roost water to feeding areas. The flights also strengthen the wing muscles of the young ducks.

November

November may begin with mild days, but by the end, subfreezing temperatures are the norm and the winter season is established. The lakes and ponds will remain frozen until spring. In some years, the region may get considerable snowfall which may not thaw until March. Prairie residents have completed preparations for the pending winter, and most of the migrants have moved on. Some Canada Geese and Mallards may stay as long as they find feed and there is open water on the rivers.

The remaining active residents also adapt to the season's change. An early snowpack may offer considerable protection for mice. A thick blanket of snow insulates the ground surface from extreme temperature change and will also provide some protection from avian predators. Snowfall also presents us with an opportunity to discover which animals are present in an area. Observing living wildlife is sometimes difficult, but they leave tracks. It can be very rewarding to venture into an area, check the tracks, and discover the activities that have taken place. It is not uncommon to identify a previously unknown resident.

Opposite page: White-tailed Deer and Mule Deer become the center of attention. November is mating time. The deer have grown their winter coats and, combined with fat reserves and hormonal changes, they look very healthy—perhaps making them attractive. The males, called bucks, forgo their secretive ways and may be seen during daylight hours. They spend time rubbing their antlers on small trees and bushes. The results are called scrapes and can readily be found. It is a sure sign that the deer are active and in the area.

Left: When protective coloration isn't! This Long-tailed Weasel has shed its brown fur of summer and its winter coat of white is in place. However, the area has not been covered in snow making the animal very easy for avian predators to see. Soon the prairie will be white, and weasels will have camouflage.

Opposite page: This Badger was looking for a meal of Prairie Dog. Badgers have very long claws and powerful front legs. They are able to hunt and catch their prey by digging into ground squirrel and prairie dog burrows. They also eat pocket gophers and a variety of ground-dwelling mammals, birds, reptiles, and amphibians. I once saw a badger carrying a cob of corn on a winter day. Perhaps it had a plan to bait rodents, or it was really hungry.

Below: Like weasels, White-tailed Jackrabbits also shed their summer fur and grow a white coat for winter. These prairie hares spend their idle time exposed to the elements and to the keen eyes of the avian predators. I can imagine that Golden Eagles would easily spot a white jackrabbit on a snowless prairie.

November is mating time for both White-tailed deer and Mule deer. Generally, during the year the bucks and does live separate lives. Bucks may seldom be seen. However, starting in early November, it is common to see bucks about during daylight hours. Their daily pattern has changed, and now activity centers on finding a mate. These three Mule Deer consist of a buck, doe, and her fawn.

This Mule Deer buck is in prime condition. During November, and the onset of mating season, hormonal changes result in the muscles bulking up to give the appearance of power. His thick neck is an obvious sign of the season. He is a handsome animal.

Left: Owls remain as year-round residents. They are able to survive cold weather and are also able to find food throughout the winter. Their diet consists of birds and mammals. Their method of attack is from a perch.

Above: On a very frosty morning, Sharp-tailed Grouse can be seen in two of the bushes. Sunrise during late fall is a good time to see coveys of sharptails in bushes and trees.

Above and left: Sharp-tailed Grouse are year-long residents of the Northern Prairie. They are well-adapted for the late fall and winter season. Beginning in late fall, flocks of sharptails can be seen in trees and bushes shortly after sunrise. They are feeding on berries from last summer and buds forming for the next growing season. They also feed on seeds and readily visit crop fields to find waste grain.

Above: Ring-necked Pheasants carry themselves with a posture of dignity. Of course, even the most dignified may at times suffer an awkward moment. The set of pheasant tracks in fresh snow reveals such a lapse. Ice had recently formed on the marsh, and perhaps the pheasant was new to the conditions. The tracks clearly show the bird was walking along, when—whoops—both feet slipped and the pheasant landed on its backside. I wonder if it looked around as we do to see if the moment was noticed.

Left: Trumpeter Swan

Opposite page: Mink are inhabitants of marsh ecosystems. Their fur stays brown all year while their weasel cousins adopt white pelts. Until the snow gets deep, mink have easy access across the marsh on the newly formed ice sheet.

By late November, the region appears void of life. Most birds have migrated, insects are gone, and amphibians and reptiles are safely underground or in marshes. Many of the mammals remain active; but some are not directly in the area.

DECEMBER

Winter is here. By the end of the month a layer of snow will cover the ground. Usually it is clean, white snow without significant drifts. Temperatures are usually tolerable for most humans, if the wind is not blowing. Only the hardy resident species, and a few that come from the north, can be seen.

Above: Food shortages compel Snowy Owls to fly to the northern prairies as winter arrives, but not every year. When rodents are abundant on the tundra, they just stay put.

Opposite page: This ancient Bur Oak on the edge of the prairie gives testimony to hardiness.

Above: I wonder if winter visitors to the Northern Prairie, such as this Lapland Longspur, think themselves to be "snowbirds" seeking warmth, just as do humans heading to Florida or Arizona.

Opposite page: Put on a heavy winter coat, you'll be fine.

Right: Clear and cold.

Opposite page: Non-native Ring-necked Pheasants bring some color to an otherwise drab landscape.

Left: Frosty mornings can betray a tawny White-tailed Deer in their normally brown habitat.

Opposite page: Another predator that follows the rodents, Short-eared Owls can reach large numbers during some winters, but disappear the next.

Left: Only the sun-dogs really celebrate on the coldest of the cold days.

Max Coleman
Age 6